BEI GRIN MACHT SICH IHR WISSEN BEZAHLT

AF135835

- Wir veröffentlichen Ihre Hausarbeit, Bachelor- und Masterarbeit

- Ihr eigenes eBook und Buch - weltweit in allen wichtigen Shops

- Verdienen Sie an jedem Verkauf

Jetzt bei www.GRIN.com hochladen und kostenlos publizieren

Bibliografische Information der Deutschen Nationalbibliothek:

Die Deutsche Bibliothek verzeichnet diese Publikation in der Deutschen National-bibliografie; detaillierte bibliografische Daten sind im Internet über http://dnb.d-nb.de/ abrufbar.

Impressum:

Copyright © 2013 GRIN Verlag
Druck und Bindung: Books on Demand GmbH, Norderstedt Germany
ISBN: 9783346078810

Dieses Buch bei GRIN:

https://www.grin.com/document/509328

Farid Badria

New Gingerol Derivative against Human Colorectal Carcinoma Caco-2 Cells

GRIN Verlag

GRIN - Your knowledge has value

Der GRIN Verlag publiziert seit 1998 wissenschaftliche Arbeiten von Studenten, Hochschullehrern und anderen Akademikern als eBook und gedrucktes Buch. Die Verlagswebsite www.grin.com ist die ideale Plattform zur Veröffentlichung von Hausarbeiten, Abschlussarbeiten, wissenschaftlichen Aufsätzen, Dissertationen und Fachbüchern.

Cytotoxicity Study of a New Gingerol Derivative and Other Related Compounds from *Zingiber officinale* against Human Colorectal Carcinoma Caco-2 cells

Pharmacognosy Department, Department of Biochemistry, Faculty of Pharmacy, Mansoura University, Mansoura 35516 Egypt

Disclosures

No conflicts of interest are declared by authors.

Key words:

Gingerols; Ginger; MPLC; Colorectal carcinoma

ABSTRACT:

In this article we report the isolation, characterization and evaluation of in vitro anti-colon cancer activity of a new gingerol derivative, namely (*S,E*)-5-hydroxy-1-(4-hydroxy-3-methoxyphenyl)tetradec-8-en-3-one, **(N6)**, and a known sesquiterpene (1,2-Dihydroxybisabola-3,10-diene,**N7**), isolated from *Zingiber officinale* rhizomes for the first time, together with other five known compounds; 6-gingerol **(N1)**, 8-gingerol **(N2)**, 10-gingerol **(N3)**, 4-gingerol **(N4)** and 4`-*O*-methyl-6-gingerol **(N5)**. The isolated compounds were identified using different spectroscopic techniques (1D and 2D-NMR, MS and IR). A modified chromatographic method was established and optimized to isolate the major gingerols **(N1, N2,** and **N3)** in grams. Thereafter the cytotoxic activity of the isolated compounds was evaluated against colorectal carcinoma cell line, Caco-2. It was observed that anti-colon cancer activity of **N2** was superior to that of 5-FU, the classic reference cytotoxic agent. Additionally, the length of the aliphatic chain in 8-gingerol **(N2)** is optimum for the anticancer activity and any decrease, as in **N1** and **N4,** or increase, as in **N3,** in the side chain length leads to gradient decrease in the cytotoxic activity. Following this further, methylation of phenolic OH group **(N5)**, leads to a dramatic decrease in the anticancer activity. Furthermore, loss of aromaticity **(N7)**, results in complete loss of the cytotoxic activity. On the other hand, introducing a π-bond in the aliphatic side chain enhances the anti-cancer activity as depicted in **N6**. Taken together, the aromaticity, the side chain length as well as the presence of free phenolic group contribute significantly to the anti-colon cancer activity of gingerol and its derivatives. These results open up a new window in the rational design of gingerol-based semisynthetic drugs with improved anticolon cancer activity.

1. Introduction

Ginger (the rhizomes of *Zingiber officinale* Roscoe), a member of the tropical and sub-tropical zingiberaceae, is globally one of the most commonly used spice. In addition to its use as a spice and condiment, ginger is also of use as a medicinal agent in the various traditional systems of medicine [1, 2]. Ginger possesses a wide array of medicinal uses and is observed to be effective against unrelated ailments. Laboratory studies have shown that ginger possesses free radical-scavenging, antioxidative, anti-inflammatory, antimicrobial, antiviral, gastroprotective, antidiabetic, antihypertensive, cardioprotective, anticancer, chemopreventive, and immunomodulatory effects [1, 2]. Currently, there is a renewed interest in ginger, and several scientific investigations aimed at isolation and identification of the active constituents of ginger and scientific verification of its pharmacological actions [1]. Ginger's numerous biological activities have been attributed mainly to gingerols and shogaols, the major pungent principles found in ginger [3, 4].

In view of the aforementioned significant bioactivities, large quantities of the pure compounds are urgently needed for further pharmacological studies. However, obtaining the pure compounds by conventional column chromatography separation methods is a challenge because of their structure similarity and unstable chemical properties [5-7]. This study develops a modified economic, rapid, and efficient method of isolation and purification to afford a high yield of gingerols with high purity to fulfill the requirement of obtaining gingerol-derived compounds. Using this optimized method, we were able to report here in the isolation of a new gingerol derivative (**N6**) from the dried rhizomes of *Zingiber officinale*, together with five known gingerols (**N1-N5**), in addition to a sesquiterpene (**N7**) which is reported for the first time from this natural source.

Extensive research over the past decade has dominated by multiple synthetic chemotherapeutic drugs that are non-invasive but display low selectivity and hence deadly adverse effects. The current cornerstone of adjuvant and palliative chemotherapy for colorectal cancer is 5-FU [9]. However, such compound shows several side effects ranging from, myelotoxicity, gastrointestinal disturbances, cardiotoxicity, and hepatotoxicity [10]. These limitations direct towards the finding of a more effective and safe drug which may raise the therapeutic benefits for patients.

Human colorectal-carcinoma is the third most common cancer diagnosed in both men and women worldwide [8]. Treatment options are limited with poor efficacy and marked patient-to-patient variation in therapeutic outcomes. In most cases, surgical resection and organ transplantation remain the only curative treatment options that may involve very expensive and invasive procedures with considerable limitations. Therefore, developing an effective therapy is of prime importance.

Chemoprevention by plant-derived compounds or dietary phytochemicals has emerged as an accessible and promising approach to cancer control and management [11-15]. Of the many phytochemicals displaying a wide array of biochemical and pharmacological activities, 6-gingerol, is the major pharmacologically active component in ginger. Recently, several lines of evidence suggest that 6-gingerol is effective in the suppression of transformation, hyperproliferation, and inflammatory processes that initiate and promote carcinogenesis, as well as the later steps of carcinogenesis, namely, angiogenesis and metastasis, and it also affect several molecular targets, and it cause modulation of several cellular pathways in the tumor cells

2

[16-18]. It was reported that 6-gingerol suppresses colon cancer growth by targeting leukotriene A_4 hydrolase [18], but there is no thoroughly study to date has profiled the anti-colon cancer activity of other gingerol derivatives. The proposed protective role of ginger and its bioactive constituents in tumor development may prevail in the intestinal tract as it has a long tradition of being very effective in alleviating symptoms of gastrointestinal problems. Preclinical studies have shown that ginger possesses carminative, gastroprotective, antiulcerative, and antiemetic properties, and it prevents epigastric discomfort, dyspepsia, stomachache, abdominal spasm, and cancer of the gastrointestinal system. It also has been recommended to combat nausea, and improve the gastrointestinal side effects associated with cancer chemotherapy [1, 19-20]. Therefore, the current study aims to evaluate the potential anti-colon cancer activity of isolated set of gingerols in comparison with 5-fluorouracil. Furthermore, this study pursues to delineate the relationships between structure elements of gingerol derivatives and their anti-colon cancer activity, in an attempt to help in a rational design of semisynthetic compounds with improved anti-colon cancer activity.

2. Results and discussion

2.1 *Chemistry*

The methanolic extract of the rhizomes of *Zingiber officinale* Roscoe was suspended in water and partitioned successively with petroleum ether, methylene chloride, ethyl acetate and n-butanol. The methylene chloride soluble fraction was subjected to a series of chromatographic techniques, leading to the isolation of a new gingerol related compound (**N6**) together with six known compounds (**N1-N5**), and **N7** which is the first time to be isolated from ginger. The known compounds were identified as 6-gingerol (**N1**), 8-gingerol (**N2**), 10-gingerol (**N3**), 4-gingerol (**N4**), 4`-*O*-methyl-6-gingerol (**N5**), and 1,2-Dihydroxybisabola-3,10-diene (**N7**), **Figure1**, by comparison of their spectroscopic data with that reported in literature [21-25]. Studies of the isolation and purification of gingerols are scarce in the literature [26]. The major gingerols (6-, 8-, and 10-gingerol) were reported to be isolated in the range of milligrams with high purity using complicated chromatographic techniques as HSCCC [5], or semi-preparative HPLC [26]. Fortunately, it is firstly reported in our study to isolate the major gingerols in the range of grams using medium pressure liquid chromatography MPLC and reversed phase C_{18} silica gel. This method can be considered as a modification for [26] new methodology for purification of gingerols using HPLC system, Luna-C_{18}, and methanol- water (75:25, v/v) as the mobile phase. In our study the best mobile phase was methanol- water (70:30, v/v) for isolation of the major gingerols, but if it is needed to isolate 4-gingerol, the best mobile phase will be methanol- water (60:40, v/v) as if methanol- water (70:30, v/v) system was used; 4-gingerol and 6-gingerol will be eluted as a mixture.

Compound **N6** was obtained as dark yellow oil. Its molecular formula is $C_{21}H_{32}O_4$ as deduced from EI-MS, $[M]^+$ peak at m/z 348.4 and $[M-H_2O]^{+\bullet}$ ion peak at m/z 330.3. The IR (neat) spectrum displayed absorption bands indicating hydroxy (3437cm^{-1}), carbonyl (1707 cm^{-1}), and aromatic (1604 cm^{-1}) functionalities. The NMR spectra of compound **N6** (table 1) is close to that of gingerols specially 10-gingerol (**N3**), as the ^{1}H-NMR spectrum of **N6** exhibited signals due to a methoxy at δ 3.80 (3H, s, H-7`) and three aromatic protons at δ 6.74 (1H, d, $J=$ 8.0 Hz, H-5`), 6.58 (2H, overlapped, H-2`,and 6`) which suggests the presence of a 4-hydroxy-3-methoxyphenyl group. This was confirmed also by the presence of a methoxy (δ 55.9, C-7`) and 6 aromatic carbons (δ 146.4, 144.0, 132.6, 120.7, 114.4, 111.0) in the ^{13}C- NMR spectrum (table 1). Furthermore, the presence of a carbonyl carbon (δ 211.3, C-3) and a hydroxymethine carbon (δ 67.2, C-5) indicated that it is a gingerol derivative. ^{13}C- NMR spectrum of compound **N6** differs than that of **N3** by the presence of only 10 aliphatic carbon signals instead of 12 carbons for the side chain and presence of two olefinic carbon signals (δ 128.7 and 130.9) indicating an unsaturation position. The presence of an extra double bond was also revealed from the ^{1}H-NMR spectrum that showed two olefinic protons at δ 5.23 (1H, overlapped dq, $J=$ 16.0, 6.4 Hz, H-8), and 5.29 (1H, overlapped dq, $J=$ 16.0, 6.4 Hz, H-9). The position of the double bond at C-8 and the assignments of the two olefinic protons were confirmed based on HSQC and HMBC spectra, which revealed that the olefinic protons, H-8 and H-9 showed HMBC correlations with aliphatic carbon signals at δ 23.3 (C-7) and 27.2 (C-10). Additional HMBC correlations (Table 1) between the protons at δ 1.30 (H-6a) and 1.41 (H-6b) with the carbons C-7, C-4, C-5, and C-8, and between the protons at δ 2.01 (H-7) with the carbons C-6, C-5,

C-8, and C-9, and between the protons at δ 1.93 (H-10) with the carbons C-11, C-12, C-8, and C-9 were used to assign the position of the double bond. Thus the structure of **N6** was established as (*S,E*)-5-hydroxy-1-(4-hydroxy-3-methoxyphenyl)tetradec-8-en-3-one, which is a new gingerol derivative.

Compound **N7** was obtained as colorless needles [from methanol- water (6:4)] with melting point 110-115 °C. Its molecular formula is $C_{15}H_{26}O_2$ as deduced from EI-MS, $[M]^+$ peak at m/z 238 and $[M-H_2O]^{+\bullet}$ ion peak at m/z 220. The IR (neat) spectrum displayed absorption bands indicating hydroxy functionality (3262 cm^{-1}). The ^{1}H-NMR spectrum (table 2) shows the presence of two olefinic protons at δ 5.04 (1H, t, *J*= 7.2 Hz, H-10) and 5.47 (1H, brs, H-4) and four methyl groups; three vinylic CH$_3$ groups at δ 1.54, 1.61, 1.74 representing protons H-13, H-12, H-15 respectively and one CH$_3$ doublet at δ 0.74 (*J*= 6.8 Hz) assigned to H-14. The two proton signals at δ 3.88 and 3.90 indicated the presence of two oxygenated methine carbons as revealed from the two carbon signals at δ 68.0 and 69.2 in the ^{13}C-NMR spectrum confirming the presence of a glycol moiety. HMBC spectrum correlations between the proton signal at δ 1.61 (H-12) with carbon signals at δ 124.6 (C-10), 131.4 (C-11), and 25.7 (C-13), the proton signal at δ 0.74 (H-14) with the carbon signals at δ 40.6 (C-6), 30.5 (C-7), and 35.2 (C-8), the proton signal at δ 1.74 (H-15) with carbon signals at δ 68.0 (C-2), 136.8 (C-3), and 130.0 (C-4), between the proton signal at δ 3.88 (H-1/2) with the carbon signals at δ 20.5 (C-15), C-4, and C-3 were used to assign the positions of the CH$_3$ groups, the two double bonds, and the glycol moiety. The chemical shift values of carbons of this compound were the same as those of the known compound 1,2-Dihydroxybisabola-3,10-diene mentioned by Gachet*et al.* 2011 [25] which is the only reference for the spectroscopic data of this compound. Although this compound was reported by Wang *et al.* 2008 [27], they did not mention its spectroscopic data. It is worth noted that there was a difference in the assignment of the proton signal at δ 1.45 (1H, dt, J = 3.6, 13.8 Hz) reported by Gachet *et al.* 2011 [25]. As they attributed this signal to H-6, while in our ^{1}H-NMR spectrum, the corresponding signal appeared at δ 1.34 (1H, dt, J= 3.2, 13.6 Hz) was assigned to H-5a and that was confirmed by the HMBC correlation from H-5a (δ 1.34) to C-7, C-6, and C-1. Neither Gachet *et al.* nor we were able to determine the relative or the absolute configuration of this compound as the crucial signals, that is, H-1/H-2 were overlapping and suitable reference data could not be found. On the basis of the spectroscopic data and comparison with literature data [25], the structure of compound **N7** was determined to be 1,2-Dihydroxybisabola-3,10-diene, which is isolated for the first time from ginger.

2.2 *Evaluation of cytotoxic activity*

The cytotoxic effects of the isolated compounds against colorectal carcinoma were evaluated in a cell-based assay using Caco-2 cells and compared to that of 5-FU, a drug extensively used in adjuvant and palliative chemotherapy for colorectal cancer. Using this approach, dose response and time course cytotoxicity of the standard agent, 5-FU, were initially carried out by MTT assay. The dose-response effect of 5-FU was more evident after 72 hours of incubation with an IC_{50} value of 60 μM than at 48 hours with IC_{50} value of 158.5 μM, whereas at 24 hours an IC_{50} was not reached with 5-FU at any concentration tested (25-250 μM). Therefore, 72 hours has been chosen as the incubation period for the dose-viability response of all tested compounds.

Generally, among the tested isolated derivatives in this cell-based assay, the two gingerols, **N2** and **N1**, showed the highest anticancer potencies of all tested derivatives that are superior to 5-FU, the classic reference cytotoxic agent. In contrast, **N4** as well as **N7**, were not active even at concentrations up to 100 μM. Interestingly, the anticancer activities of **N5**, **N6** and **N3** had moderate IC_{50} values that were substantially lower than that of **N4** and **N7** but higher than those of **N2** and **N1**.

Moving over the molecular structure of the isolated derivatives, it was observed that the length of the aliphatic chain in **N2** is optimal for anti-colon cancer activity that showed the highest anticancer potency superior to 5-FU, **Figure 2A**. Any decrease in the side chain length causes either a significant decrease in the cytotoxic activity as depicted in **Figure 2B** for **N1** (2-carbons less) or even totally loss of anticancer activity as depicted in **Figure 2C**, for **N4** (4-carbons less). Likewise, the increase in the side chain length leads to dramatic decrease in the cytotoxic activity, as shown in compound **N3**, **Figure 2D**. However, it was found that the presence of π-bond in the aliphatic chain as in compound **N6** increases the cytotoxic activity compared to compound **N3** which has the same structure but differs from **N6** in the absence of the double bond in the side chain, **Figure 3A.**

The methylation of the phenolic OH group resulted in a substantial decrease in anti-colon cancer activity as illustrated by **N5**, (4`-*O*-methyl-6-gingerol) which showed 50% inhibition of cell viability after 72 hr. incubation at concentration of 65.73 ± 2.5 μM, **Figure 3B**, in comparison with (6-gingerol, **N1** withIC_{50}= 37.7 μM), indicating the importance of free phenolic OH group in the anti-colon cancer activity of gingerols. Following this further, loss of aromaticity has a greater influence on the anti-colon cancer activity of gingerols. This is clearly manifested in **N7** in which the cytotoxic activity has been dramatically diminished over the used concentration range, **Figure 3C**.

One-way ANOVA was used to test for statistical differences in anti-colon cancer effect at IC_{50} for each compound after 72 hr. incubation period, and the means were significantly different across the samples, **Figure 4**.

3. Conclusions

In conclusion, with the continuing need for novel drug-like lead compounds against the increasing number of ever-more challenging molecular cancer targets, the chemical diversity derived from natural products will be increasingly relevant to the future of drug discovery. Therefore, the activity of the studied naturally derived series from *Zingiber officinale* rhizomes in inhibiting colon carcinoma cells make them potential leads for anti-colon cancer agents.

We have isolated and characterized a new gingerol derivative (**N6**), together with other five known gingerols, and a sesquiterpene isolated for the first time from ginger (**N7**). We were able to isolate the major gingerols in grams, using a modified chromatographic method including medium pressure liquid chromatography (MPLC), and reversed phase C_{18} silica gel. The isolated compounds were evaluated for their cytotoxic activity towards colorectal carcinoma. The results revealed that the length of the aliphatic chain, the presence of unsaturation in the aliphatic chain, the aromaticity, and the presence of free phenolic OH group greatly affect the cytotoxic activity of gingerols.

4. Experimental

4.1. General experimental information

Melting points were determined on Stuart® melting point apparatus model SMP10 and are uncorrected. ^{1}H and ^{13}C-NMR spectra were obtained in $CDCl_3$ solutions with TMS as internal standard at 400 MHz for ^{1}H-NMR and 100 MHz for ^{13}C-NMR on JEOL Eclipse NMR or BRUKER Ascend™ 400 spectrometer. Chemical shifts (δ) are reported in ppm relative to the solvent signal and coupling constants are given in Hz. Mass spectrometry (EI-MS) spectra were obtained using a Thermo Scientific DSQ™ II instrument. IR spectra were obtained using a Thermo Scientific Nicolet™ iS™10 FT-IR spectrometer instrument. Column chromatography was carried out on silica gel G 60-230 mesh (Merck, Germany) and reversed phase C_{18} silica gel (BAKERBOND® octadecyl C_{18}, 40μm). Analytical thin layer chromatography (TLC) is performed on precoated silica gel 60 GF_{254} (20 x 20 cm, 0.2 mm thick) on aluminium or plastic sheets (Merck, Germany), and Partisil® KC18F Silica gel 60A with fluorescent indicator (5 x 20 cm, 200μm layer thickness) Whatman®. Whereas UV and vanillin/sulfuric acid spray reagent were used as revealing agents. Organic solvents were distilled prior use.

4.2. Plant material

The dry rhizomes of *Zingiber officinale* Roscoe were purchased at a herbal market in Mansoura, Egypt in March 2011. A voucher specimen was deposited in the herbarium of the college of pharmacy, Mansoura University (010-Mansoura-1).

4.3.1. Extraction and isolation

The powdered rhizomes of *Zingiber officinale* Roscoe (3kg) were extracted by maceration in a glass jar with distilled methanol (5 x 4 L) at room temperature. The combined methanol extract was then concentrated under vaccum at 45 C° then allowed to dry to a constant weight to obtain a crude methanolic extract (280.3 g). The dried alcoholic extract was dissolved in a small amount of methanol and diluted with the same volume of distilled water in a separating funnel. Using liquid-liquid partitioning method; it was extracted successively till exhaustion with petroleum ether, methylene chloride, ethyl acetate and n-butanol. The fractions, in each case, were evaporated under vacuum at 45 C° to afford petroleum ether fraction (113.4 g, 40.5%), methylene chloride fraction (64.77 g, 23.1%), ethyl acetate fraction (21 g, 7.5%), n-butanol fraction (6 g, 2.1%), and aqueous fraction, which was not processed.

It was found that only the petroleum ether and methylene chloride fractions were containing gingerols. However, the methylene chloride fraction was used in gingerols' isolation due to its higher content of gingerols and lesser amount of oil compared to the petroleum ether fraction.

The methylene chloride fraction (60 g) was subjected to normal phase silica gel (350 g) column chromatography (CC) packed in petroleum ether (100%) and successive gradient elution with petroleum ether- ethyl acetate. Based on TLC analysis, similar fractions were pooled together to afford 7 major fractions (containing gingerols), eluted with the solvent system petroleum ether- ethyl acetate (9:1). These fractions (**Figure 5**) can be classified according to their major contents into; fraction G1 (1.54 g, contains major compound **N3**), fraction G2 (1.35 g, contains compounds **N3** and **N2**), fraction G3 (5.34 g, contains compounds **N3**, **N2**, and **N1**), fraction G4 (9.70 g, contains compounds **N2**, and **N1**), fraction G5 (3.83 g, contains major compound **N1**), fraction G6 (1.60 g, contains compounds **N1**, and **N4**), and fraction G7 (1.50 g, contains major compound **N4**).

Fraction G3 (5 g) was rechromatographed over reversed phase C_{18} silica gel (about 200 g) using medium pressure liquid chromatographic column (MPLC) and eluted with isocratic solvent system of methanol- water (7:3) to afford compound **N1** [1.7 g, pale yellow oily liquid, R_f0.62 using C_{18} silica gel TLC and methanol- water (8:2) as eluting system, quenched UV_{254} light and gave orange color with vanillin/sulfuric acid spray reagent after heating at 110 °C for 1 min], compound **N5** [28 mg, pale yellow oily semisolid, R_f0.53using C_{18} silica gel TLC and methanol- water (8:2) as eluting system, quenched UV_{254} light and gave orange color with vanillin/sulfuric acid spray reagent after heating at 110 °C for 1 min], compound **N2** [0.8 g, pale yellow oily liquid, R_f0.45 using C_{18} silica gel TLC and methanol- water (8:2) as eluting system, quenched UV_{254} light and gave orange color with vanillin/sulfuric acid spray reagent after heating at 110 °C for 1 min], compound **N6** [15 mg, dark yellow oily liquid, R_f0.40 using C_{18} silica gel TLC and methanol- water (8:2) as eluting system, quenched UV_{254} light and exhibited a yellow fluorescence under UV_{336} light and gave orange color with vanillin/sulfuric acid spray reagent after heating at 110 °C for 1 min], and compound **N3** [1.2 g, pale yellow powder, R_f0.30 using C_{18} silica gel TLC and methanol- water (8:2) as eluting system, quenched UV_{254} light and gave orange color with vanillin/sulfuric acid spray reagent after heating at 110 °C for 1 min].

Fraction G7 (1.5 g) was also rechromatographed over reversed phase C_{18} silica gel CC (20 cm x 2.5 cm, about 100 g) but no pressure was applied and eluted with isocratic system of methanol- water (6:4) to afford compound **N4** [0.19 g, pale yellow oily semisolid, R_f0.72 using C_{18} silica gel TLC and methanol- water (8:2) as eluting system, quenched UV_{254} light and gave orange color with vanillin/sulfuric acid spray reagent after heating at 110 °C for 1 min], compound **N7** (**Figure 6**) which was obtained by crystallization from methanol-water (6:4), [15 mg, colorless fine needles, R_f0.25 using normal silica gel TLC and petroleum ether- ethyl acetate (2:1) as eluting system, showed no behavior under UV light, gave gray color with vanillin/sulfuric acid spray reagent after heating at 110 °C for 1 min]. In addition, compound **N1** (73 mg) was isolated again from this column.

4.3.2. Characterization

1,2-Dihydroxybisabola-3,10-diene (**N7**): colorless fine needles, (15 mg); IR (neat) v_{max} 3262, 2959, 2931, 2873, 2855, 1674, 1647, 1604, 1516, 1452, 1376, 1269, 1065, 1029, 987, 884, 818, 737 cm^{-1}; ^{13}C-NMR (CDCl$_3$, 100 MHz) δ 136.8 (C-3), 131.4 (C-11), 130.0 (C-4), 124.6 (C-10), 69.2(C-1), 68.0 (C-2), 40.6 (C-6), 35.2 (C-8), 30.5(C-7), 29.8 (C-5), 26.1 (C-9), 25.7 (C-12), 20.5 (C-15), 17.7 (C-13), 14.4 (C-14); EI-MS C$_{15}$H$_{26}$O$_2$, peak at m/z 238 [M]$^+$, peak at m/z 220 [M-H$_2$O]$^{+\bullet}$, peak at m/z 202 [M-2H$_2$O]$^{+\bullet}$, other fragments appeared at m/z 153, 135, 121, 109, 81, and 55.

4.4 Cytotoxicity assay

Materials

Dulbecco's Modified Eagle's Medium (DMEM) was obtained from Gibco Laboratories (Life Technologies Inc., Grand Island, NY, USA). Non-essential amino acids (NEAA) and fetal calf serum (FCS) were purchased from Hyclone (Logan, UT, USA). Penicillin and streptomycin were purchased from Amresco (Solon, OH, USA). Plates of 96 wells were purchased from Corning Costar (Cambridge, MA, USA).

Cell culture

Human colon adenocarcinoma cell line, Caco-2 (ATCC #HTB-37) was purchased from American Type Culture Collection (ATCC, Rockville, MD, USA). Caco-2 cells were cultured in DMEM containing D-glucose (4.5 g/L), NaHCO$_3$ (3.7 g/L), supplemented with 10% FCS, penicillin (100 U/mL) and streptomycin (100 μg/mL) in an atmosphere of 5% CO$_2$ and 90% relative humidity at 37 °C. All cells used in this study were between passages 50 and 62.

Drugs treatment. The concentrations of the studied isolated compounds were in the range from 25 to 250 μM. The drugs were dissolved in 100% dimethylsulfoxide (DMSO; Sigma) and then diluted in the media for experiments. In all experiments, control cells were incubated with DMSO alone. The final concentration of DMSO was maintained at 0.2%. Cells were incubated with tested compounds or 5-FU for 72 h.

MTT assay. This assay relies on the ability of viable cells to reduce a yellow tetrazolium salt (MTT; Sigma) metabolically to a purple formazan product [28-29]. This reaction takes place when mitochondrial reductase enzymes are active. Cells were grown in 96-well plates ($1 \times 10^4/200$ μL/well). After incubation with the reagents, the medium was removed and the cells were treated with 75 μL of MTT and the cells were further incubated for 3 hr. at 37 °C. Subsequently, 100 μL of DMSO were added to each well to dissolve the crystals of the viable cells, and the solubilized formazan product was spectrophotometrically quantified with the help of a microtiter plate reader, Power Wave XS (Bio Tek, Winooski, VT, USA), at 540 nm. The cytotoxic activity of the compounds was indicated by the percentage cell viability which was calculated with the following formula: % cell viability = (average absorbance of treated cells/ average absorbance of control cells) x100 and IC_{50} was calculated.

Statistical analysis. Data were presented as mean values ± standard deviation (SD). Statistical comparisons between groups were performed by one-way analysis of variance (ANOVA) followed by Tukey's post hoc test (Statistica, Stat Soft, USA). Value of $p < 0.05$ was assumed as statistically significant.

Table 1. ^{1}H-, ^{13}C- NMR and HMBC spectroscopic data for compound **N6**.

Position	^{1}H-NMR[a]	^{13}C-NMR[a]	HMBC[b]
1	2.75 (2H, t, J= 7.2 Hz)	29.3	2, 2`, 6`, 1`, 3
2	2.64 (2H, t, J= 7.2 Hz)	45.5	1, 4, 1`, 3
3	---	211.3	
4	Ha: 2.40 (1H, overlapped dd, J= 10.0, 18.4 Hz) Hb: 2.48 (1H, overlapped)	49.4	6, 2, 5, 3
5	3.98 (1H, brs)	67.2	7, 6
6	Ha: 1.30 (1H, m) Hb: 1.41 (1H, m)	36.3	7, 4, 5, 8
7	2.01 (2H, m)	23.3	6, 5, 8, 9
8	5.23 (1H, overlapped dq, J= 16.0, 6.4 Hz)	128.7	7, 6, 9, 10
9	5.29 (1H, overlapped dq, J= 16.0, 6.4 Hz)	130.9	7, 11, 8
10	1.93 (2H, q, J= 6.4 Hz)	27.2	11, 12, 8, 9
11	1.21-1.28 (6H, m)	29.4	10, 12, 13
12	1.21-1.28 (6H, m)	31.5	14, 13, 11, 10
13	1.21-1.28 (6H, m)	22.6	14, 12,11
14	0.80 (3H, t, J= 6.4 Hz)	14.1	13, 12
1`	---	132.6	
2`	6.58 (2H, overlapped)	111.0	1, 6`, 1`, 4`, 3`
3`	---	146.4	
4`	---	144.0	
5`	6.74 (1H, d, J= 8.0 Hz)	114.4	6`, 1`, 4`, 3`
6`	6.58 (2H, overlapped)	120.7	1, 2`, 4`, 3`
7`	3.80 (3H, s)	55.9	3`

[a]In CDCl$_3$, at 400 MHz for ^{1}H and 100 MHz for ^{13}C. Coupling constants (J) are in Hz.

^b HMBC (Heteronuclear multiple bond correlation) correlations are from proton(s) stated to the indicated carbon.

Table 2. ^{1}H-NMR[a] spectroscopic data for compound **N7**.

position	N7	1,2-Dihydroxybisabola-3,10-diene[*]
1	3.88 (1H, overlapped)	3.96 (1H, overlapped)
2	3.90 (1H, overlapped)	3.97 (1H, overlapped)
3	---	---
4	5.47 (1H, brs)	5.54 (1H, s)
5	5a: 1.34 (1H, dt, J= 3.2, 13.6 Hz)	1.45 (1H, dt, J = 3.6, 13.8 Hz, H-6)
	5b: 1.64 (1H, overlapped d, J= 14.0 Hz)	1.73 (1H, m, H-5)
6	1.54 (1H, overlapped)	1.65 (1H, overlapped, H-6)
7	1.91 (1H, m)	1.99 (1H, m)
8	1.23 (2H, q, J= 7.2 Hz)	1.33 (2H, overlapped)
9	1.94 (2H, overlapped)	2.03 (2H, overlapped)
10	5.04 (1H,t, J= 7.2 Hz)	5.13 (1H, br t, J = 6.0 Hz)
11	---	---
12	1.61 (3H, s)	1.68 (3H, s)
13	1.54 (3H, s)	1.61 (3H, s)
14	0.74 (3H, d, J= 6.8 Hz)	0.82 (3H, d, J = 7.2 Hz)
15	1.74 (3H, s)	1.81 (3H, s)

[a]In CDCl$_3$, at 400 MHz, Coupling constants (J) are in Hz.

[*]^{1}H- NMR data for1,2-Dihydroxybisabola-3,10-diene [Gachet *et al.* 2011].

References:

[1]B.H. Ali, G. Blunden, M.O. Tanira, A. Nemmar, Food Chem. Toxic. 46 (2008) 409–420.

[2] A.R Shivashankara, R. Haniadka, R. Fayad, P.L. Palatty, R. Arora, M.S. Baliga, Elsevier Inc. (2013) 657- 671.

[3] J.J. Araya, H. Zhang, T.E. Prisinzano, L.A. Mitscher, B.N. Timmermann, phytochem.72 (2011) 935-941.

[4] M. Afzal, D. Al-Hadidi, M. Menon, J. Pesek, M.S.Dhami, Drug Metabol. Drug Interact. 18 (2001) 159–190.

[5] X. Wang, Z. Zheng, X. Guo, J. Yuan,C. Zheng,Food Chem. 125 (2011) 1476–1480.

[6] R. Grzanna, L. Lindmark, C.G. Frondoza, J. Med. Food 8 (2005) 125–132.

[7] H.A. Schwertner, D.C. Rios, J. Chromatogr. B 856 (2007) 41–47.

[8] R.T. Greenlee, M.B. Hill-Harmon, T. Murray, M. Thun, CA Cancer J. Clin. 51 (2001) 15-36.

[9] L. Lombardi, F. Morelli, S. Cinieri, D. Santini, N. Silvestris, N. Fazio, L. Orlando, G. Tonini, G. Colucci, E. Maiello, Cancer Treat. Rev.36S3 (2010) S34 –S41.

[10] Meta-Analysis Group in Cancer, J. Clin. Oncol. 16 (1998) 3537-3541.

[11] F.A. Badria, Cancer Lett. 84 (1994) 1-5.

[12] F.A. Badria, M.Ameen, M.R. Akl, J.Biosci.62 (2007) 656-660.

[13] A.A. Fadda, F.A. Badria, K.M. El-Attar, Med. Chem. Res.19 (2010) 413-430.

[14] W.M. Alarif, A. Abdel-Lateff, S.S. Al-Lihaibi, S.N. Ayyad, F.A. Badria, J. Z. Naturforsch. 68 (2013) 70 – 75.

[15] Y.J. Surh, Nat. Rev. Cancer 3 (2003) 768–80.

[16] A.M. Bode, Z. Dong, The Amazing and Mighty Ginger, in: I.F.F. Benzie, S. Wachtel-Galor, Herbal Medicine: Biomolecular and Clinical Aspects, 2nd edition, Boca Raton (FL): CRC Press, (2011) 131-156.

[17] Y. Shukla, M. Singh, Food Chem. Toxicol. 45 (2007) 683–690.

[18] C.H. Jeong, A.M. Bode, A. Pugliese, *et al.*, Cancer Res. 69 (2009) 5584–5591.

[19] S. Chrubasik, M.H. Pittler, B.D. Roufogalis, J. Phytomed. 12 (2005) 684–701.

[20] M.S. Baliga, A.R. Shivashankara, R. Haniadka, P.L. Palatty, R. Arora, R. Fayad, Elsevier Inc. (2013) 187-199.

[21] N. Shoji, A. Iwasa, T. Takemoto, Y. Ishida, Y. Ohizumi, J. Pharm. Sci. 71 (1982) 1174–1175.

[22] S.W. Lee, J. Lim, M.S. Kim, J. Jeong, G. Song, W.S. Lee, M. Rho, Food Chem. 128 (2011) 778–782.

[23] J.S. Kim, S.I. Lee, H.W. Park, J.H. Yang, T.Y. Shin, Y.C. Kim, Arch. Pharm. Res. 31 (2008) 415–418.

[24] H. Kikuzaki, Y. Kawasaki, N. Nakatani, J. Am. Chem. Soc. 24 (1994) 237–243.

[25] M.S. Gachet, O. Kunert, M. Kaiser, Brun R., Zehl M., Keller W., R.A. Munoz, R. Bauer, W. Schuehly, J. Nat. Prod. 74 (2011) 559–566.

[26] J.A.d. Silva, A.B. Becceneri, H.S. Mutti, A.C.B.M. Martin, M.F.G.F. Silva, J.B. Fernandes, P.C. Vieira, M.R. Cominetti, J. Chromatogr. B 903 (2012) 157–162.

[27] L.Y. Wang, M. Zhang, C.F. Zhang, Z.T. Wang, Acta Pharm. Sin. 43 (2008) 724–727.

[28] T. Mosmann, J. Immunol. Methods 65 (1983) 55–63.

[29] R.F. Hussain, A.M.E. Nouri, R.T.D. Oliver, J. Immunol. Methods 160 (1993) 89 –96.

FIGURE LEGENDS

Figure 1. Structural characteristics of the isolated compounds from *Zingiber officinale* Rhizomes.

Figure 2. Effect of the side chain length on the cytotoxic activity of isolated gingerols: (A) IC_{50} of the most potent compound (8-gingerol, **N2**). (B & C) IC_{50} of 6-gingerol (**N1**) & 4-gingerol (**N4**), that is shorter in side chain length than **N2**. (D) IC_{50} of 10-gingerol (**N3**) that is longer in side chain length than **N2**. These experiments have been done after 72 hr. incubation with different concentration of the isolated compounds (25, 100, and 200 μM), counted as percentage of untreated control, 0 μM). Data shown are the mean ± SD of three experiments.

Figure 3. Effect of introducing π-bond in the aliphatic side chain, methylation of aromatic OH group, or loss of aromaticity on the cytotoxic activity of isolated gingerols: (A) IC_{50} of the new gingerol derivative (**N6**) that has a π-bond in the aliphatic side chain compared to **N3**.

(B) IC_{50} of 4`-*O*-methyl-6-gingerol (**N5**) that has a methylated aromatic OH group in **N1**. (C) IC_{50} of compound **N7**, (without aromatic ring). These experiments have been done after 72 hr. incubation with different concentration of the isolated compounds (25, 100, and 200 μM), counted as percentage of untreated control, 0 μM). Data shown are the mean ± SD of three experiments

Figure 4. Statistical differences in anti-colon cancer effect at IC_{50} for each isolated phytochemical after 72 hr. incubation period, and the means were significantly different across the samples. Data shown are the mean ± SD of three experiments.

Figure 5. C_{18} silica gel TLC using methanol-water (8:2) as eluting system and visualization using vanillin/sulfuric acid spray reagent after heating at 110 °C for 1 min. (A) for methylene chloride column fractions isolated by normal phase silica gel CC; (B) & (C) for compounds **N5**, **N6**, and **N3** isolated from fraction G3 using reversed phase C_{18} silica gel CC.

Figure 6. Crystallization of compound **N7** from methanol-water (6:4).

Figure 1

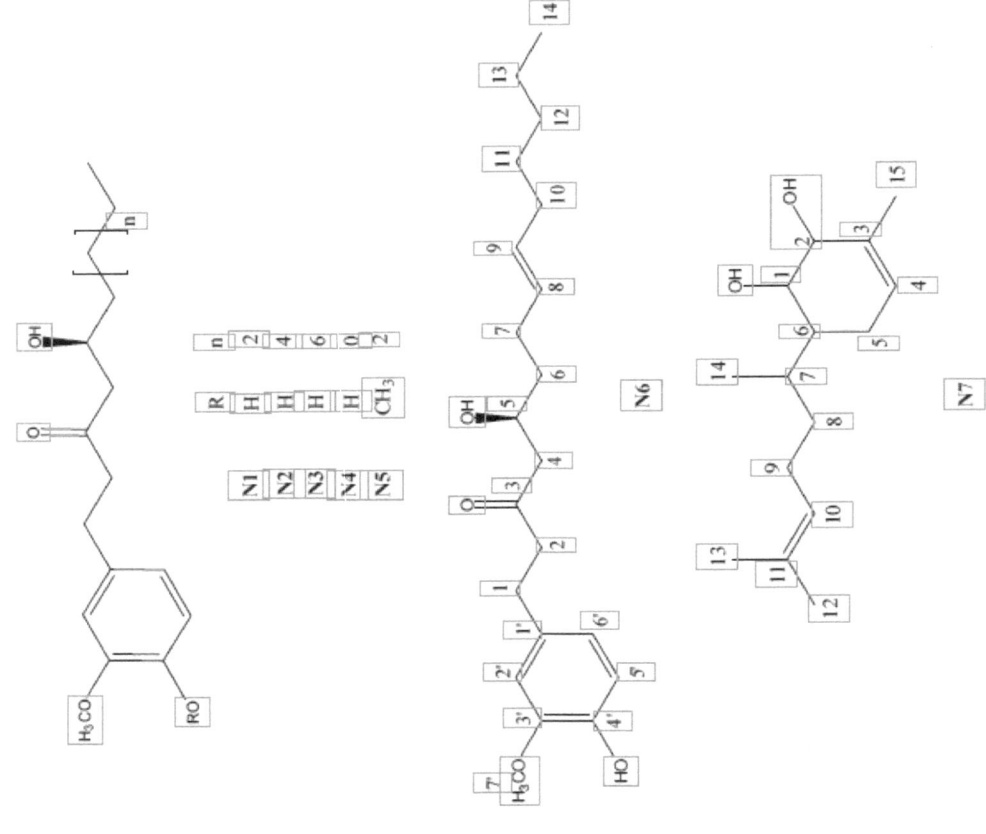

Figure 2

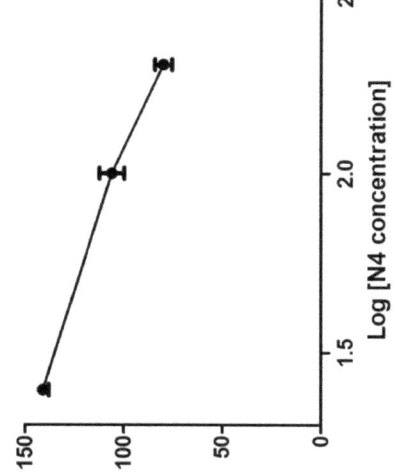

C

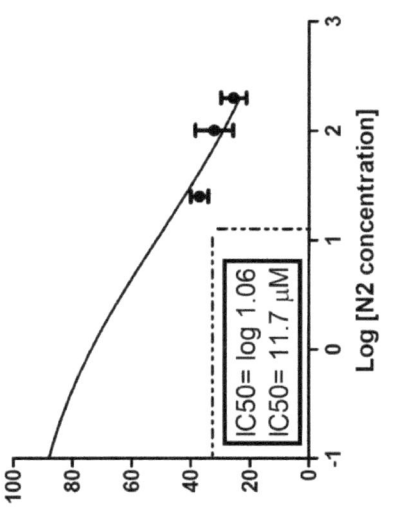

A

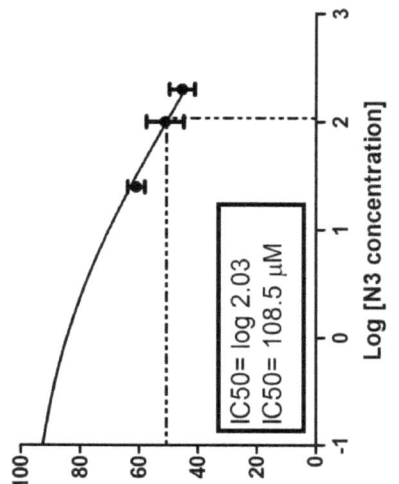

D

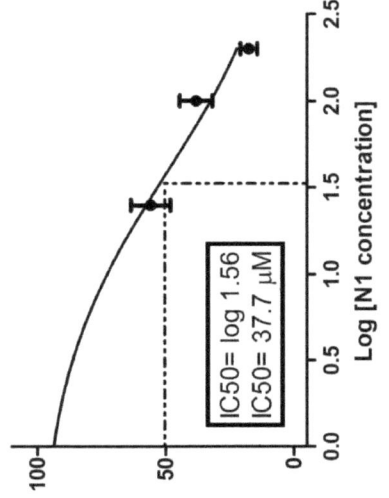

B

Figure 3

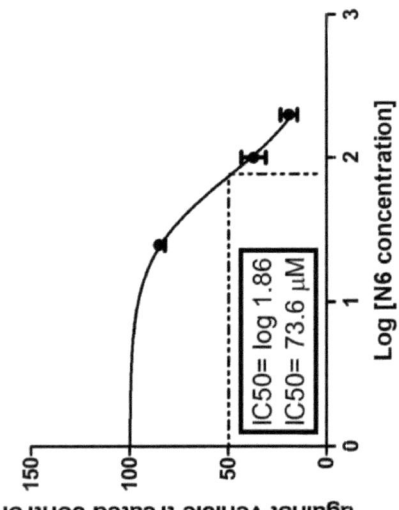

A

% of Cell viability normalised against vehicle treated control

IC50= log 1.86
IC50= 73.6 μM

Log [N6 concentration]

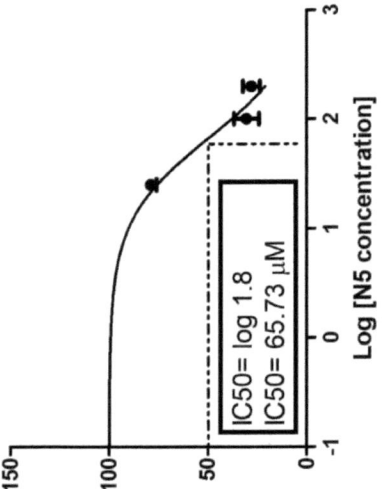

B

% of Cell viability normalised against vehicle treated control

IC50= log 1.8
IC50= 65.73 μM

Log [N5 concentration]

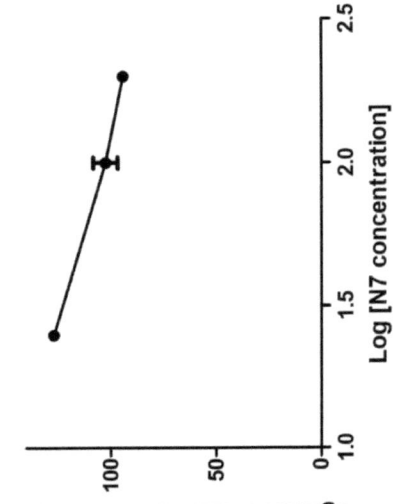

C

% of Cell viability normalised against vehicle treated control

Log [N7 concentration]

Figure 4

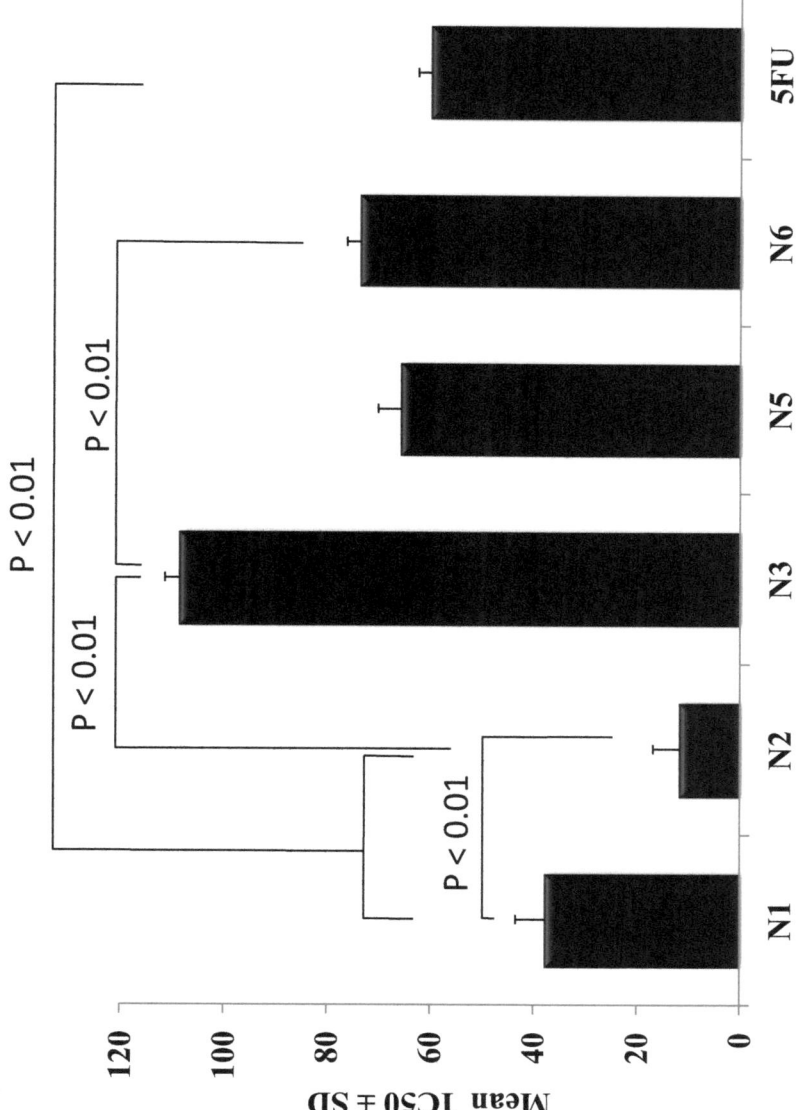

Figure 5

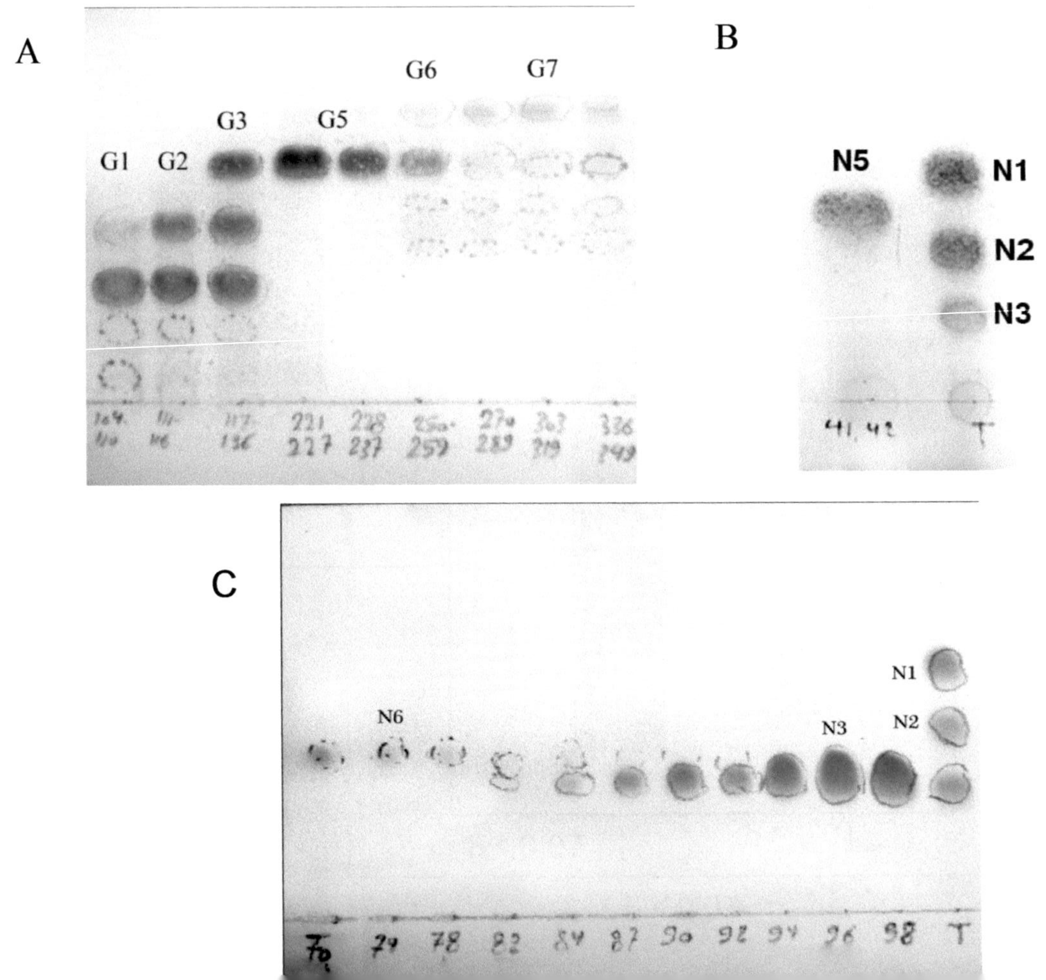

A

B

C

N1
N2
N3
N5

N1
N2
N3
N6

Figure 6

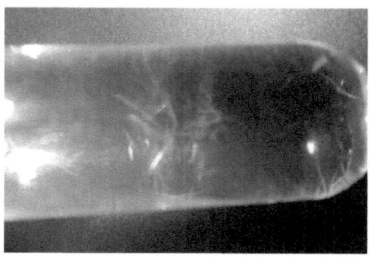

BEI GRIN MACHT SICH IHR WISSEN BEZAHLT

- Wir veröffentlichen Ihre Hausarbeit,
  Bachelor- und Masterarbeit

- Ihr eigenes eBook und Buch -
  weltweit in allen wichtigen Shops

- Verdienen Sie an jedem Verkauf

Jetzt bei www.GRIN.com hochladen und kostenlos publizieren